Power PRACTICE™

MW01678721

Applying Problem-Solving Strategies

3–4

Written by

Angela Higgs

Editor: Heather Butler
Illustrator: Jenny Campbell
Production: Karen Nguyen
Cover Designer: Barbara Peterson
Art Director: Moonhee Pak
Project Manager: Collene Dobelmann
Project Director: Betsy Morris

Table of Contents

Introduction

Applying Problem-Solving Strategies 3–4 contains dozens of ready-to-use activity pages to provide students with skill practice. The fun activities can be used to supplement and enhance what you are already teaching in your classroom. Give an activity page to students as independent class work, or send the pages home as homework to reinforce skills taught in class. An answer key is provided at the end of the book for quick reference.

This book provides activities that will directly assist students in practicing strategies needed to solve word problems. The activities are grouped in three-page sets that cover each problem-solving strategy. The first activity page of each set includes the subhead "Show Me the Way" because it demonstrates step-by-step for the student how to use the strategy. The remaining pages offer students a chance to practice using the strategy to solve similar problems. Please note that the problems in each section could be solved using multiple strategies.

Use these ready-to-go activities to "recharge" skill review and give students the power to succeed!

Name Working Backwards Date ____________

drew a picture
excluded info

Homework Time

SHOW ME THE WAY TO WORK BACKWARDS

~~Before dinner, Cira did her homework.~~ She spent 15 minutes practicing her spelling words. Right after that she spent twice as long completing her math homework. She finished at 4:50 p.m. What time did she start practicing her spelling words? ____________

Strategic Steps

1 If you work backwards, you can easily find Cira's start time. She spent 15 minutes practicing her spelling and twice as long on her math homework. Using addition, double the time spent on spelling to find out how much time she spent on math homework.

15 minutes + 15 minutes = 30 (time spent on math homework)

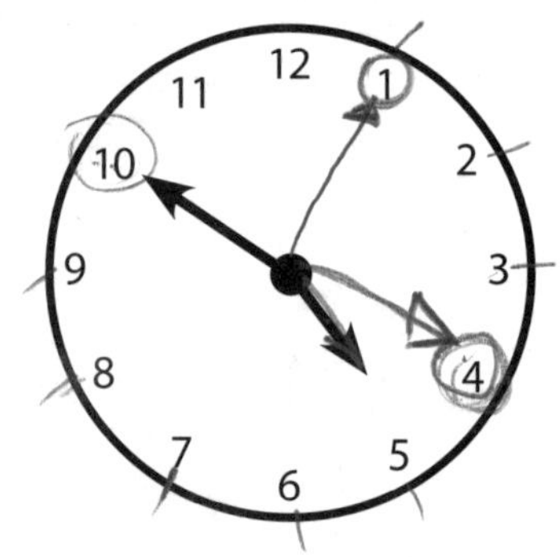

2 Then start at 4:50 p.m. Subtract the time she spent on math homework. 4:20

3 Next, subtract 15 minutes from this new time to find out what time she started practicing her spelling words. 4:05 She started her Spelling at 4:05.

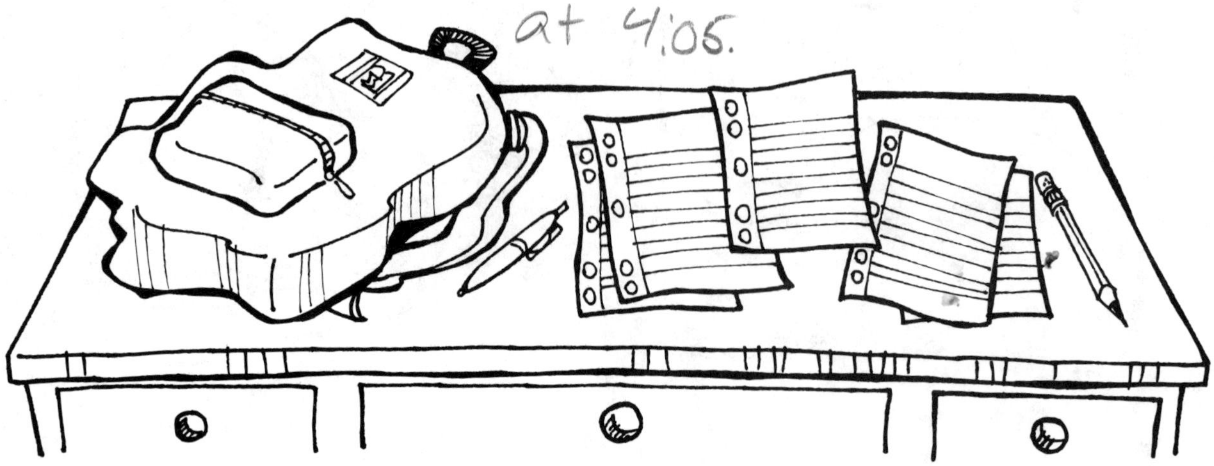

Name ______________________________ Date ______________

Backwards Time

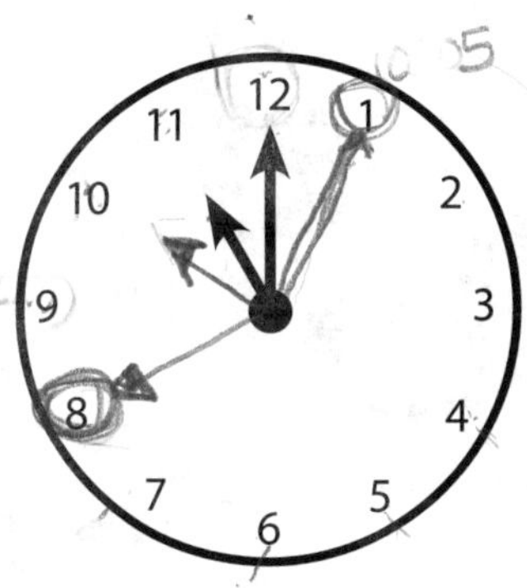

A Hunter has several chores to do before he can go swimming with his friend. Unloading the dishwasher and putting away the dishes takes 25 minutes. He can dust and vacuum the house in 40 minutes. Cleaning his room takes 15 minutes. If Hunter's friend picks him up at 11:00 a.m., what is the latest time he can start his chores? 1 hr 20 min

9:40 am

11:00
− 1 hr. 20 min
9:40

B Arthur needs to arrive at school at 8:00 a.m. It takes him 10 minutes to shower, 5 minutes to get dressed, 15 minutes to eat breakfast, 5 minutes to finish getting ready and get his things together, and 10 minutes to walk to school. What time must he get up to arrive at school on time?

~~8:15~~
7:15

C Jenny wants to watch a movie on TV that starts at 7:00 p.m. First she has to clear the table, which takes 10 minutes, and spend the next 25 minutes finishing her homework. Then she reads to her sister for a half hour. What time does she have to start clearing the table so that she can watch the movie? She has to Start Clearing the table at 5:30

Name ______________________________ Date ______________

Hobbies

A Jasper made cookies to share with his friends. He put all of the cookies on a plate and passed it around. Johnny took 6 cookies, Brandy took half as many cookies as Johnny, Clarissa took 2 more than Brandy, and there were 3 cookies left. How many cookies did Jasper make?

____

B Vanessa rode her bicycle from school to the library, from the library to the park, and then home. She rode a total of 9 miles. If it is 2 miles from the library to the park and 3 miles from the park to her house, how far is it from the school to the library?

____

C Diamond loves to go to the library and goes once a week. The first week she checked out 5 books, the second week she checked out twice as many books, and the third week she checked out 7 books. If she checked out 26 books in January, how many books did she check out the last week?

____

Name ______________________________ Date ______________

Air Hockey Tournament

Show Me the Way to Solve a Simpler Problem

Four players are participating in an air hockey tournament. Each of these players must play each other. How many games will be played?

____

Strategic Steps

1. Break this problem into several small steps. Assign each of the 4 players a letter to make it easier, such as A, B, C, and D. Start with player A and match that player with all the other players.

Player **A**: A vs. ______________

A vs. ______________

A vs. ______________

2. Move on to the other players and continue listing all the new games each player will play. Remember not to list the same game twice. For example, A vs. B and B vs. A is the same game and should not be repeated.

Player **B**: B vs. ______________

B vs. ______________

Player **C**: C vs. ______________

Player **D**: no new games

3. Once you have finished, count the number of games to find the total played during the tournament.

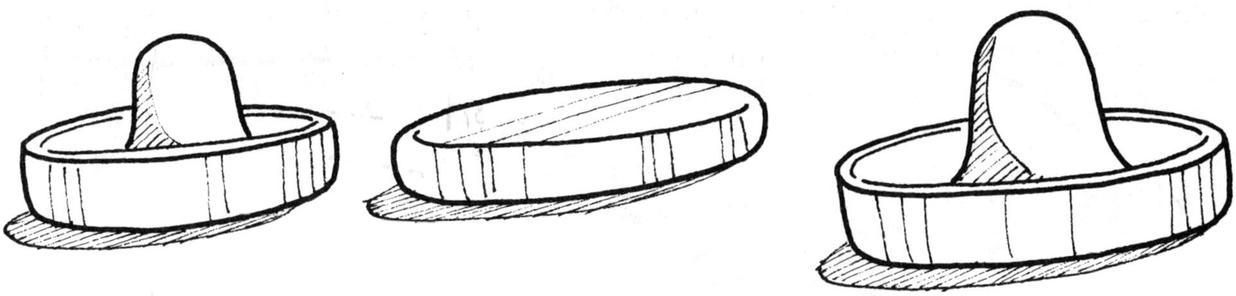

Name ______________________ Date ____________

Geography Bee

Mrs. Postelle's class is having a geography bee. There are 6 finalists. The finalists will be paired with each of the other finalists to answer a different question. The other class members will write the questions.

How many questions will the class need to write? ________

Finalist **A**: ________ ________ ________ ________ ________

Finalist **B**: ________ ________ ________ ________

Finalist **C**: ________ ________ ________

Finalist **D**: ________ ________

Finalist **E**: ________

Finalist **F**: ________

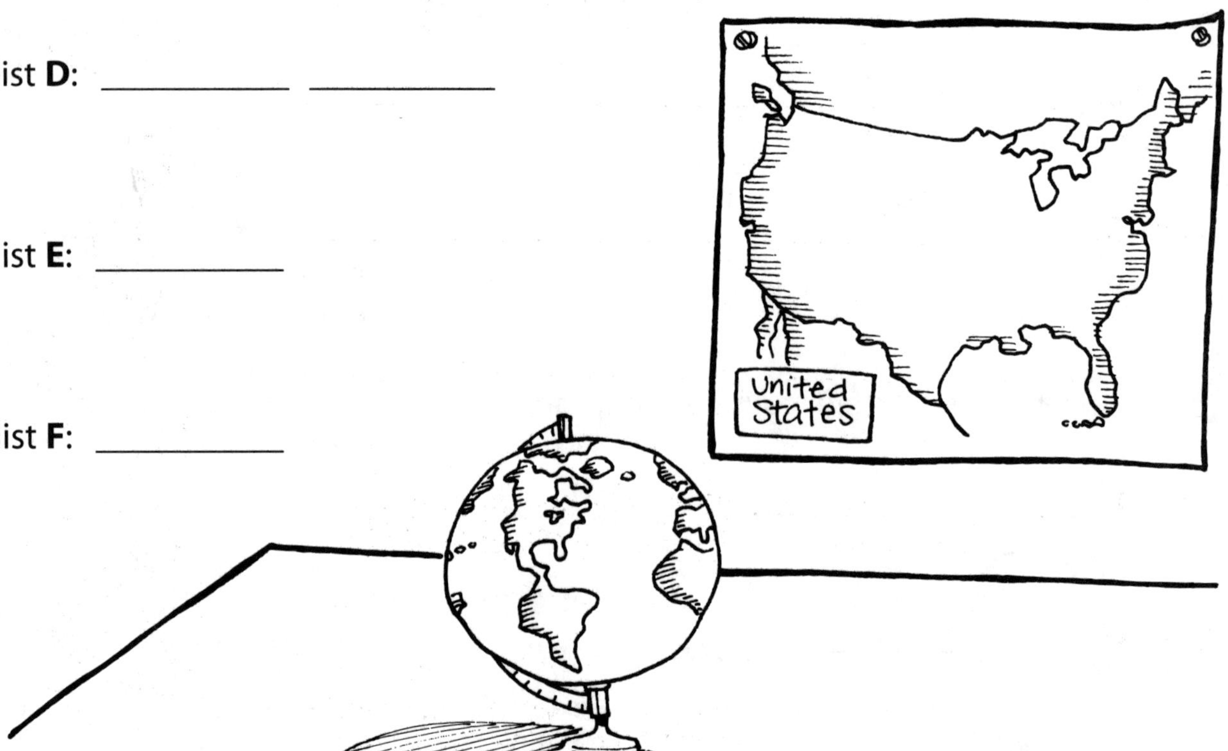

Name ______________________________ Date ____________

Game Time

Mrs. Colley's class is having a board game tournament. Eight students have signed up to play in the tournament. In each game, one student will be paired against another. The games will continue until every student has played each of the other seven students.

How many games will be played in the tournament? ____________

Player **A**: ______________________________

____

Player **B**: ______________________________

____

Player **C**: ______________________________

Player **D**: ______________________________

Player **E**: ______________________________

Player **F**: ______________

Player **G**: ______________

Player **H**: ______________

Name ______________________________ Date ______________

A Penny Saved

Show Me the Way to Solve Multiple-Step Problems

A John earns $10 each week by mowing the lawn for his dad. He gets the same amount in allowance each week. Every week he goes to the arcade and spends $4. He puts the rest of his money in his piggy bank. After six weeks, how much money will he have saved?

____

Strategic Steps

1. First, find out how much money John has coming in each week. Add the money he earns to the amount of his allowance. ________ + ________ = ________
2. Next, see how much he saved each week. Take the total from Step 1 and subtract from it the money he spent at the arcade. ________ – $4 = ________
3. Then take this amount and multiply it by 6 to see how much he saves over six weeks. ________ × 6 = ________

B Melissa is starting a jewelry business. She wants to make beaded necklaces, bracelets, and earrings to sell. To start her business, she needs to buy supplies. She has $100 to spend. Assorted beads cost $45, the wire costs $8, and other supplies cost $23. How much money will she have left after buying all the supplies?

____

Strategic Steps

1. First, add up the cost of her supplies.

   ________ (beads) + ________ (wire) + ________ (other supplies) = ________

2. Next, subtract the total of Step 1 from the $100 she has to spend.

   $100 – ________ (answer from Step 1) = ________ (how much she'll have left)

Name ______________________________ Date ______________

The Pet Store

A Gerry is going to buy fish for his new aquarium. He can purchase a total of 23 fish. He picks out 8 guppies and 9 mollies. He wants to buy tetras to finish filling the aquarium. How many tetras can he buy?

____

B Annie needs to buy food for her two kittens. Together the kittens eat a small can of cat food three times a day. How many cans of cat food will Annie need to feed the kittens for one week?

____

C Don and Joey's mom is letting them buy two cockatiels. The cockatiels are $62 each, the cage costs $50, and food costs $8. The boys have saved $200. How much money will they have left after buying the birds and supplies?

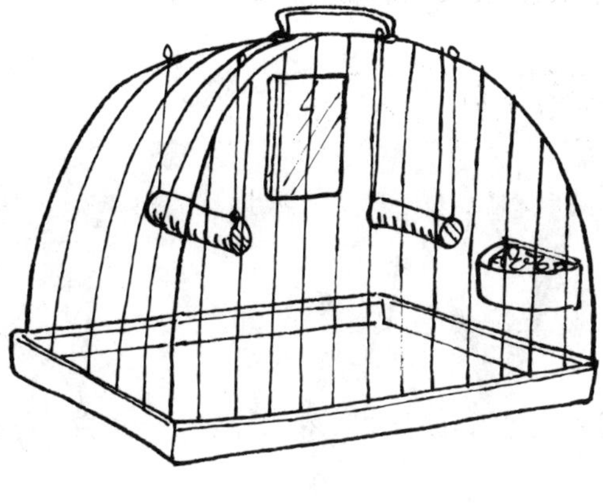

____

Name ______________________________ Date ____________

Insects

Mrs. Gilley is teaching her class about insects. She made "insects" from different colors of pipe cleaner and hid them in the school yard. The students were directed to find as many pipe-cleaner insects as they could, group them by color, and count the number in each group.

The chart below shows the number of pipe-cleaner insects the students found.

Blue	8
Green	5
Red	12
Black	8
Brown	3

Mrs. Gilley hid 53 insects. Did the students find them all? ____________

If not, how many more need to be found? ____________

Name ____________________________ Date ____________

Field Trip

SHOW ME THE WAY TO EXCLUDE EXTRA INFORMATION

Ms. Moore's class went on a field trip to the natural history museum. All 22 students in her class went on the field trip along with 5 parents who went as chaperones. Mrs. Chong's class went on the same bus. She had 21 students and 6 parent chaperones. At the museum, they met up with Mr. Adams's class of 24 students and 4 parent chaperones.

How many students were on the bus with Ms. Moore? ____________

Strategic Steps

1. There is a lot of information in this problem. Some of it is not needed to solve the problem. What is the problem asking you to find out?

   ____

2. Do you need to know how many parent chaperones went with each class? ____________

3. Were all the students on the same bus? ____________

4. The answers from Steps 2 and 3 were both *no*, so just add together the number of students from the two classes that were on the same bus as Ms. Moore.

________ (Ms. Moore's students) + ________ (Mrs. Chong's students) = ________

Name ______________________________ Date ______________

Rock Quarry

A Mr. Fulton's class went on a field trip to the rock quarry. This quarry was really fun because they could find all sorts of rocks and bring them home. Mike found a four-pound milky quartz rock. Arley found two pink quartz rocks. One was three pounds and the other was seven pounds. Kathryn found a smoky quartz rock that weighed twelve pounds. Bryan found two feldspar rocks—one pink feldspar that weighed two pounds and one green feldspar that weighed one pound. Kathryn found a piece of granite that weighed nine pounds.

How many pounds of rocks were quartz? ______________

B Some of Mr. Fulton's students were fascinated with the rock quarry. After the field trip, eight children said they would like to become geologists. Four students did not like the quarry and said they were bored. Five students said they would like to become teachers one day so that they could plan field trips to the rock quarry.

How many students wanted to become teachers or geologists? ______________

Name ______________________________ Date ______________

The Fishing Trip

A Valerie's family went on a fishing trip. They drove 156 miles to the first river to go fishing. There they caught 18 fish. They ate 3 of the fish for lunch. At dinner they ate another 2 fish. The next morning they drove 38 miles to another river. There they caught 15 more fish. They ate 4 fish for lunch. Then they drove 120 miles home and had friends over for dinner. They cooked 10 fish. They froze the rest of the fish.

How many miles did Valerie's family drive? ______________

B Valerie wrote in her journal during her fishing trip. She kept track of the fish her family caught. At the first river they caught 4 salmon, 6 rainbow trout, 3 white fish, 2 brown trout, and 3 perch. In the other river they caught 3 bass, 2 bluegill, 6 rainbow trout, and 4 brook trout. She also wrote about when she saw a river otter catch 2 rainbow trout.

How many trout did her family catch altogether? ______________

Name ______________________________ Date ______________

Cardinals' Season

Show Me the Way to Use Data from a Chart

The following chart shows the statistics from a 12-game season of the Cardinals' Little League baseball team.

Player	Hits	Singles	Doubles	Triples	Home Runs
Melissa	10	5	3	0	2
Christopher	15	6	6	2	1
Liz	18	11	4	1	2
Sammie	23	12	5	1	5
Katie	8	1	6	0	1
Nick	11	2	5	2	2
Ron	15	6	5	1	3
David	10	2	4	1	3
Jeff	9	2	5	1	1
Kayla	18	8	5	1	4

A How many hits were single-base hits? ______________

Strategic Step

1. Look at the chart and add all the numbers under the "Singles" column.

5 + 6 + 11 + 12 + 1 + 2 + 6 + 2 + 2 + 8 = ______________

B What is the difference between the greatest number of hits by a player and the fewest number of hits by a player?

____

Strategic Step

1. On the chart, find the player with the most hits. Then find the player with the fewest. Subtract the fewest number of hits from the greatest number of hits.

______________ − ______________ = ______________

greatest fewest

Name ______________________ Date ____________

Reading Contest

One of Mrs. Evan's reading groups made a chart to keep track of the number of pages they read each week during a reading contest. Use this chart to answer the questions below.

Student	Week 1	Week 2	Week 3	Week 4	Total
Savanna	312	78	56	35	481
Doug	98	135	672	173	1,078
Gerald	315	278	346	24	963
Jayson	102	114	99	121	436

A How many total pages did the children read altogether?

B How many more pages did Jayson read during week 4 than week 3?

C What is the difference between the greatest and the fewest total number of pages read?

Name ______________________________ Date ______________

Pet Parade

The third grade took a survey to find the number and kinds of pets the students owned. They entered the data in the following chart.

Class	Dogs	Cats	Fish	Birds	Other Pets
Mrs. Estrada's	7	5	2	1	2
Mr. Sutherland's	6	9	4	0	4
Mrs. Mullen's	8	2	1	2	3
Ms. Dee's	9	3	1	2	1

A Which class has the most pets? ______________________

B In Ms. Dee's class, how many more cats are there than birds? ______________

C How many dogs do the third graders have altogether? ______________

D What is the fraction comparing the number of dogs owned by Mrs. Estrada's students to the total number of dogs owned by third graders?

____

Name ______________________________ Date ______________

Movie Treats

Show Me the Way to Guess and Check

Lisa went to the movies and spent $4.00 on snacks. Popcorn was $2.50, a candy bar was $1.25, a soda was $2.75, and mints were $1.75. Which two snacks did she buy?

____

Strategic Steps

1. Make a guess and then check to see if the answer is correct. We know that the two snacks need to add up to $4.00, so only one of them can cost over $2.00. The first guess could be popcorn and mints.

   $2.50 + $1.75 = ______________

2. That equals more than she paid, so one item must be less. The second guess could be popcorn and a candy bar.

   $2.50 + $1.25 = ______________

3. That equals less than she paid. Which two items did she buy? Show your work.

   $ ______________ + $ ______________ = $4.00

Name ______________________ Date ____________

Make Your Best Guess

A Shannon and Kate were counting the pencils in their pencil boxes. Shannon had 4 more pencils than Kate. Together, they had 20 pencils. How many pencils does each girl have?

B Tom and Gordon collect baseball cards. Tom has 12 more cards than Gordon. Together, they have 74 cards. How many cards does each boy own?

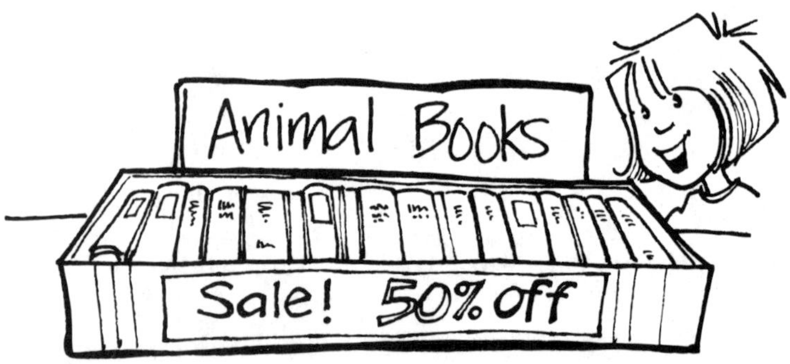

C Amanda bought two books at the book fair. She spent a total of $10.00. One book was a dollar more than the other. How much did each book cost?

D Miranda and Becca went jogging. Miranda got tired and stopped. Becca jogged for another 5 minutes. Altogether, they jogged for 35 minutes. How many minutes did each girl jog?

Name ______________________________ Date ______________

School Olympics

A Roger and Xavier are practicing for the long jump. Xavier jumped a foot farther than Roger. Altogether, they jumped 9 feet. How far did each boy jump?

B Brenna and Neha like to run sprints and time each other with a stopwatch. Neha's time was 3 seconds more than Brenna's time. If you add the girls' times together, you get 19 seconds. How long did it take each girl to run the race?

C Edward and Tyler were practicing for the discus throw. Tyler threw the discus 5 feet farther than Edward. Altogether, they threw the discus 25 feet. How far did each boy throw the discus?

D Mike and Annabelle were competing in the jump rope contest. Annabelle jumped 9 more times than Mike. Altogether, they jumped 43 times. How many times did each child jump?

Name ______________________________ Date ____________

Number Changing Box

Show Me the Way to Look for a Pattern

Mrs. Denkers showed her class a box and told them that it was a NCB—a Number Changing Box. She explained that if you wrote a number on a piece of paper and put it in the box, when you pulled the paper out of the box, the number would change. The class's job was to figure out what the box did to change the number. Mrs. Denkers told the class that the NCB would add, subtract, divide, or multiply by a certain number to change the number. She told them that they should write down all the numbers going into the box and all of the numbers coming out to be able to figure out the pattern, which is called the **function**.

Elke put in the number 6 and was given the number 8. Alexis put in the number 15 and was given the number 17. Wali said, "I know what is happening! I bet if I enter the number 5, a 7 will come out." Which operation and number did the Number Changing Box use to determine the function?

____

Strategic Steps

1. Did the numbers increase or decrease when they came out of the box? ________
(**Hint:** If they increased, then the operation is addition or multiplication. If they decreased, then subtraction or division was used.)

2. In the first blank in the number sentences below, fill in the operation you think the NCB used. In the second blank, write the same number in each of the remaining blanks that goes with the operation to make each number sentence true. If each number sentence is true and has the same operation and number in each blank, then you have determined the correct function. If not, try a different operation or number.

	In	Operation	Number	Out
Elke	6	____________	____________	= 8
Alexis	15	____________	____________	= 17
Wali	5	____________	____________	= 7

Name ______________________________ Date ______________

What Happened?

A Mrs. Denkers's Number Changing Box adds, subtracts, divides, or multiplies by certain numbers to change numbers. Help her class figure out the operation and number of each function the NCB is using.

These are the numbers that went in and came out of the box:

In	Out
40	24
66	50
57	41
161	145

Which operation and number did the NCB use to complete the function? ______________

B Mrs. Denkers shook the Number Changing Box to change its function.

These are the numbers that went in and came out of the box:

In	Out
3	9
12	36
20	60

Which operation and number did the NCB use to complete the function? ______________

C You found out that you can control the NCB. What operation and number do you want the NCB to use to complete the function? ______________

Write some numbers that would work with the function you chose.

In	Out
________	________
________	________
________	________
________	________

Name ______________________ Date ____________

My Number Is Different!

A The NCB is at it again! Only this time it's using two functions to change a number. Help Mrs. Denkers's class discover the operation and number of the functions. These are the numbers that went in and came out of the box:

In	Out
10	21
12	25
15	31
22	45

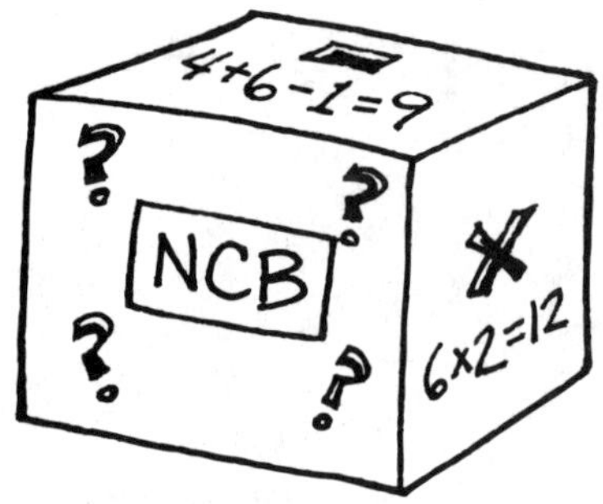

Which operation and number did the NCB use to complete the functions? ____________

B The NCB got knocked off the table. The students put the same numbers in and different numbers came out. It's still using two functions to change each number. These are the numbers that went in and came out of the box:

In	Out
10	32
12	38
15	47
22	68

Which operation and number did the NCB use to complete the functions? ____________

C You are in control of the NCB. What operation and number do you want the NCB to use to complete the function? ____________
Write some numbers that would work with the function you chose.

In	Out
________	________
________	________
________	________
________	________

Name ______________________________ Date ______________

Flower Garden

SHOW ME THE WAY TO MAKE A CHART

Nina, Sue, and Inna wanted to plant a flower garden at the school. They decided that on Monday they would plant 3 flowers altogether. Then each day during the week they would plant twice as many flowers as the day before.

A How many flowers will they plant on Friday? ______________

Strategic Steps

1. We know that Nina, Sue, and Inna were going to plant 3 flowers on Monday so we can start a chart with that information and then add to it all week.

Day	Flowers Planted
Monday	3
Tuesday	
Wednesday	
Thursday	
Friday	

2. The girls were going to plant twice as many flowers on Tuesday as they did on Monday. So if you multiply the number from Monday by 2, you will find how many flowers they will plant on Tuesday.

$3 \times 2 =$ ______________ (number of flowers planted on Tuesday)

3. Continue to fill in the chart until you find how many flowers were planted on Friday.

B How many total flowers will they plant that week? ______________

Strategic Step

1. To find the total number of flowers the girls planted for the week, add the numbers in the "Flowers Planted" column.

3	+	______	+	______	+	______	+	______	=	______
Monday		Tuesday		Wednesday		Thursday		Friday		Total for week

Name ______________________________ Date ______________

Pennies

Brynn's mom gave her a penny. Then her mom told her that she will give her twice as many pennies as the day before for the next nine days. Make a chart showing how many pennies Brynn will get each day.

What was the total amount of money Brynn got from her mom? ______________

Name ______________________________ Date ______________

Pebbles

Mike, Decker, and Sirius were collecting pebbles to add to the water garden being built at the neighborhood park. On Monday they collected 5 pebbles and took them to Mr. White, the gardener. Mr. White told them that they needed to collect twice as many pebbles each day for the next 7 days to have enough for the water garden. Make a chart to see how many pebbles the boys collected each day.

What is the total number of pebbles the boys collected? ______________

Name ______________________________ Date ______________

Lunch Line

SHOW ME THE WAY TO DRAW A PICTURE

Grant, Ethan, Ben, Emma, and Lauren were in line for lunch. Here are some clues to help you find the order in which they were standing:

- Ethan is first in line.
- Grant is fourth in line.
- Ben is between Grant and Ethan.
- Grant is in front of Lauren.
- Emma is front of Ben.

Strategic Steps

1. The first two clues tell us where Ethan and Grant are, so label them in the picture.

2. The third clue says that Ben is between Grant and Ethan. Right now there are two places between Grant and Ethan, so move on to the next clue and come back to this one later.

3. The fourth clue tells us that Grant is in front of Lauren. Since there is only one place after Grant, that means that Lauren is the last in line. Fill in her name on the picture.

4. That leaves the third and fifth clues to tell us where Ben and Emma are standing. Use these clues to label Ben's and Emma's location. Make sure Emma is in front of Ben.

Name ______________________________ Date ______________

Back to School

This year Tara, Jordan, Lance, Mia, and Nora are assigned to different classrooms. Their classrooms, rooms 8, 9, 10, 11, and 12, are numbered in order going down the hall.

Here are some clues to which classrooms they are in:

- Tara's classroom is between Jordan's and Lance's classrooms.
- Lance is in the classroom next to Mia.
- There are three classrooms between Nora's and Mia's.
- Mia's class is in the first classroom, room 8.

Write each child's name next to the classroom he or she is in.

Room 8 ______________________ Room 11 ______________________

Room 9 ______________________ Room 12 ______________________

Room 10 ______________________

Name ______________________________ Date ______________

Science Notebooks

For science, each child has a notebook with his or her name on it. Juan, Leah, Rita, Miguel, Susan, and David are working at the same lab table and have placed their notebooks in one stack.

Here are some clues to help you figure out the order of the six notebooks:

- There is one notebook between Miguel's and Leah's.
- Miguel's notebook is between David's and Susan's notebooks.
- There aren't any notebooks on top of Rita's.
- Leah's notebook is touching the table.
- David's notebook is two under Rita's.

What order from top to bottom are the six notebooks in?

____

____

Name ______________________________ Date ______________

The Bridge

Show Me the Way to Solve Multiple-Answer Problems

Devin was lying on the grass near a bridge. Some people crossed the bridge walking their dogs. When he started counting, he counted 10 legs in the group that crossed. What combinations of dogs and people could have been in that group? Write all the different combinations:

____

Strategic Steps

1 Let's see if we can find at least two answers to this problem. An easy way to do this is to use a chart for organization.

People	Legs	Dogs	Legs	Total Number of Legs	Does it Work?
1	2	2	8	10	Yes
2	4	1	4	8	No

2 You have found one answer that works. The second combination listed does not work because the total number of legs does not add up to 10. Continue trying other combinations until you find another combination that has a total of 10 legs as the answer.

Name ______________________________ Date ______________

One Potato, Two Potatoes

Vicki bought 50 pounds of potatoes. The potatoes were sold in 5-pound bags and 10-pound bags. Fill in the chart with all the different combinations of 5-pound bags and 10-pound bags that she could have bought.

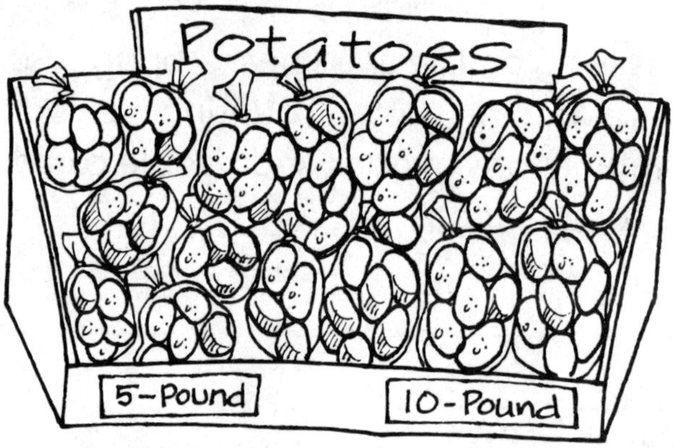

Number of 5-Pound Bags	Number of 10-Pound Bags	Does it Work?

Name ______________________________ Date ______________

The Picnic

Roy and Brent went on a picnic with their families. Roy started a game and told Brent that he counted 24 legs. Some were ants' legs and some were people's legs. Roy wanted Brent to guess all of the different combination of ants and people that this could be. Fill in the chart to find all the different combinations.

People	Legs	Ants	Legs	Total Number of Legs	Does it Work?

Name ______________________________ Date ______________

Close Enough

SHOW ME THE WAY TO ESTIMATE

Sometimes you need an exact answer to a problem, but other times you just need to be close to the answer. Read each of these problems and decide if you need an exact answer or an estimate.

A The school record in the jump roping contest is 87 jumps without missing. You lost track when you were jumping and thought you'd jumped about 90 times. If you want to beat the record, do you need to start over and count the exact number of jumps or can you go with your estimate? Explain.

____

Strategic Steps

1. Does it matter if you have an exact number or can you just be close to the number?

____

2. In this case, you must know the exact number of times you jumped to compare to the school record. An estimate wouldn't let you know if you set a new record.

B You and a friend are playing computer games. You have 128 points and your friend has 43 points, about how many more points do you have than your friend?

____

Strategic Steps

1. Since the problem uses the word *about,* you can make an estimate instead of finding the exact amount. Round each point total to the nearest ten.

Actual point total	Rounded to the nearest ten
128	______________
43	______________

2. Use mental math to subtract 40 from 130 since you have rounded off to the nearest tens.

Name ______________________________ Date ______________

Estimation

A Your family is planning a trip this weekend. You would like to go to the amusement park that is 127 miles away from home. Your sister wants to go hiking at a state park that is 72 miles away. About how much farther from your house is the amusement park than the state park?

B You are running a sprint. The school record is 14.1 seconds. Your time is 14.4 seconds. Can you round the numbers to the nearest second and say you tied the school record? Explain.

____

____

C Emily and Rodney were comparing their keychain collections. Rodney has 33 keychains on his backpack, and Emily only has 12 keychains. About how many more keychains does Emily need in order to have as many as Rodney?

Name ______________________________ Date ______________

Friends

A Emily and her friend Roberto both collect baseball cards. Emily has 174 cards. Roberto has 291 cards. About how many more cards does Roberto have than Emily?

B It is 10:00 a.m., and Sam wants to meet his friends to go sledding. First, he has to shovel the snow from his sidewalk. Last time there was this much snow it took him 18 minutes to shovel the sidewalk. It will take him about 8 minutes to walk to the sledding hill. About what time should he tell his friends he can meet them?

C The school record for using the same brown paper lunch bag is 93 days. Donald has used his lunch bag for 36 days. If he rounds each of these numbers to the nearest ten to find out about how many more days he needs to beat the record, will this work? Explain.

____

____

Name ______________________ Date ____________

Going to School

SHOW ME THE WAY TO FIND ASSUMED STEPS

Ruth Ann always walks to school. She walks a mile and a half to school. How many miles does she walk roundtrip?

____

How many miles does she walk in a regular school week? ______________

Strategic Steps

1. This is a tricky problem because it is asking for information that is assumed. This means that some of the information needed is not stated in the problem. For example, the problem doesn't say how many days you go to school in a week; it assumes that you already know that a regular school week is five days.

2. This problem also has a second assumed step. It states that Ruth Ann walks to school and lives a mile and a half from school. It does not say that she walks home, but the word "roundtrip" is used. This word tells you that if she walks to school, she will also walk home. That doubles the distance she walks in a day.

3. She walks a mile and a half to school; therefore, to get home, she walks a mile and a half again. Add the two distances together to find how many miles she walks in a day.

   1 ½ miles + 1 ½ miles = ______________ miles walked in a day

4. How many miles does she walk in a week? You calculated the number of miles she walks in a day and know that she goes to school five days a week.

   ______________ miles walked in a day × 5 days = ______________ miles walked in a school week

Name ______________________________ Date ______________

Paper Drive

A The members of Naomi's Earth Club are collecting newspapers as a fund-raiser. Their goal is to collect 50 pounds of newspaper every day after school for a week. If they reach their goal every day, how many pounds will they collect in a week?

B Naomi's club did not meet its goal of 50 pounds a day for a week. Instead the members only collected 200 pounds for the week. If they continue to collect this same amount each week, how many pounds will they collect in one month?

C If they started collecting in September and collected 1,000 pounds of newspaper each month, how many pounds will they have collected by New Year's Day?

Name ________________________________ Date ________________

Horseback Riding

A Alyssa went to a horseback riding camp on her summer vacation. She loved spending time grooming and feeding the horses. Her favorite horse, Snickers, ate 5 pounds of grain per day. How many pounds of grain did Snickers eat in one week?

B Alyssa's favorite time of day was when she would practice jumping hurdles with Snickers. Every day they practiced for 2 hours. How many hours did they practice in the month of July?

C Alyssa rode through the pastures on horseback with her new friends. They rode 8 miles each morning and 6 miles each afternoon. How many miles did they ride during the month of June?

Name ______________________________ Date ______________

Venn Diagram

Show Me the Way to Use Venn Diagrams

There are 14 children in the scout troop. They are working on a pet care badge and were given either a goldfish, a mouse, or both to take care of. Three children got only a goldfish. Five children have both a goldfish and a mouse. How many got only a mouse?

Strategic Steps

1. Use a Venn diagram to organize the data. Label one circle *Goldfish* and the other circle *Mouse*. The area where the two circles overlap is for children who have both a goldfish and a mouse.

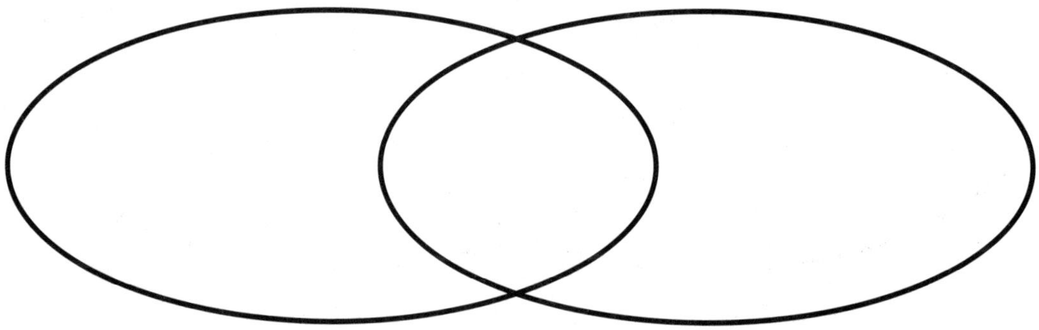

______________________ ______________________

2. Next, fill in the information that you have been given. In the Goldfish circle, write the number of children who have a goldfish. In the overlapping section write the number of children who have both a goldfish and a mouse. The Mouse section will be empty for now.

3. Find how many children have only a mouse. First, add together the number of children who have just a goldfish and the children who have both a goldfish and a mouse.

________	+	________	=	________
Children with just a goldfish		Children with a goldfish and mouse		

4. Now, subtract the answer from Step 3 from the total number of children in the scout troop.

________	–	________	=	________
Total children		Sum from above		Children with just a mouse

Name ______________________________ Date ______________

Play Time

A There are 23 students in Mrs. Pop's third-grade class at Farmington Elementary School. She asked the students to decide if they liked soccer or baseball more. Thirteen students chose soccer and seven students chose baseball. The rest of the students liked both sports. How many students chose both soccer and baseball? Fill in the Venn diagram to help you solve this problem.

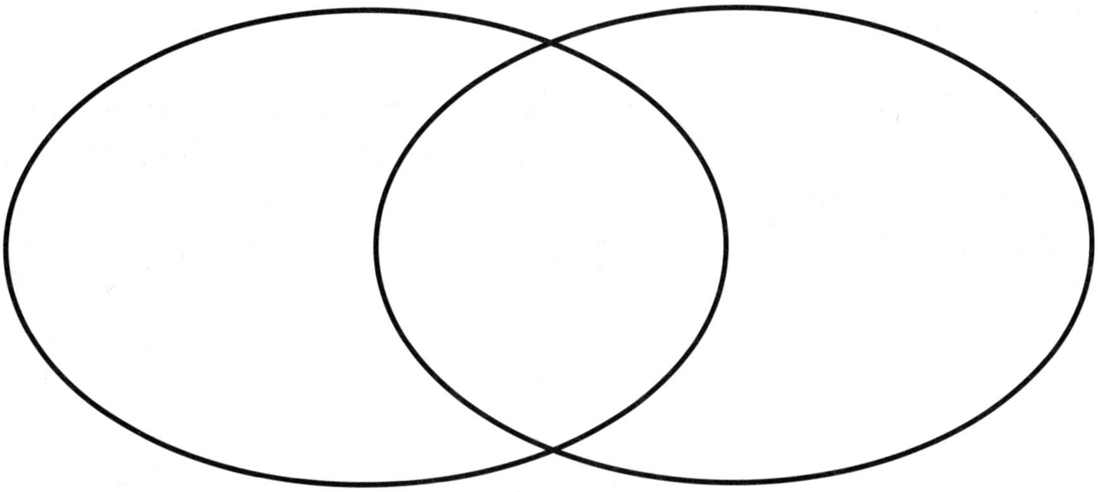

Soccer **Baseball**

B The PTO raised enough money to add new playground equipment to the playground. The 36 fourth-grade students voted on which equipment to add: 12 voted to add just a climbing wall, but 14 wanted both a climbing wall and a new set of monkey bars. The rest wanted only new monkey bars. How many students wanted just monkey bars? Fill in the Venn diagram to help you solve this problem.

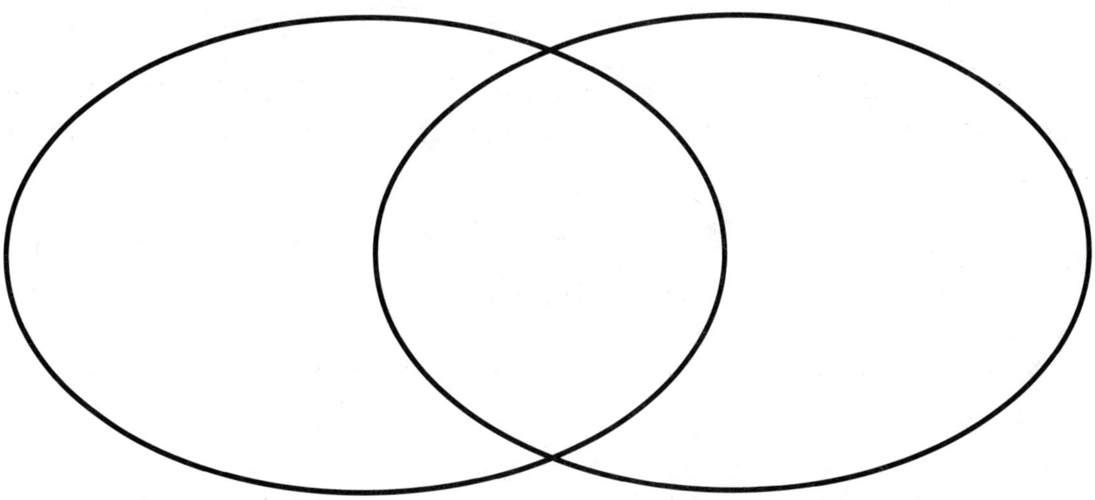

Climbing Wall **Monkey Bars**

Name ______________________________ Date ______________

Survey Says

A Kate took a survey of third graders to find out who preferred to take vacations in the mountains or at the beach. She found that 11 students liked just the beach, 8 students liked just the mountains, and 6 students liked both. How many students did she survey? Fill in the Venn diagram to help you solve this problem.

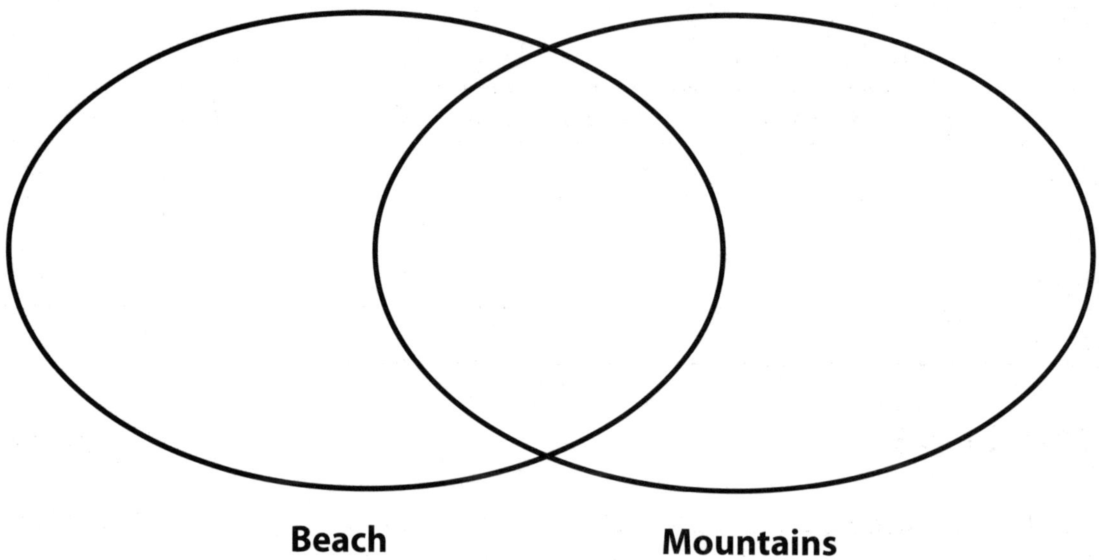

B Roger took a survey of fourth graders to find out how many liked card games or board games. He talked to 24 students and found that 6 liked only card games, 5 liked only board games, and the rest liked both card and board games. How many children liked both types of games? Fill in the Venn diagram to help you solve this problem.

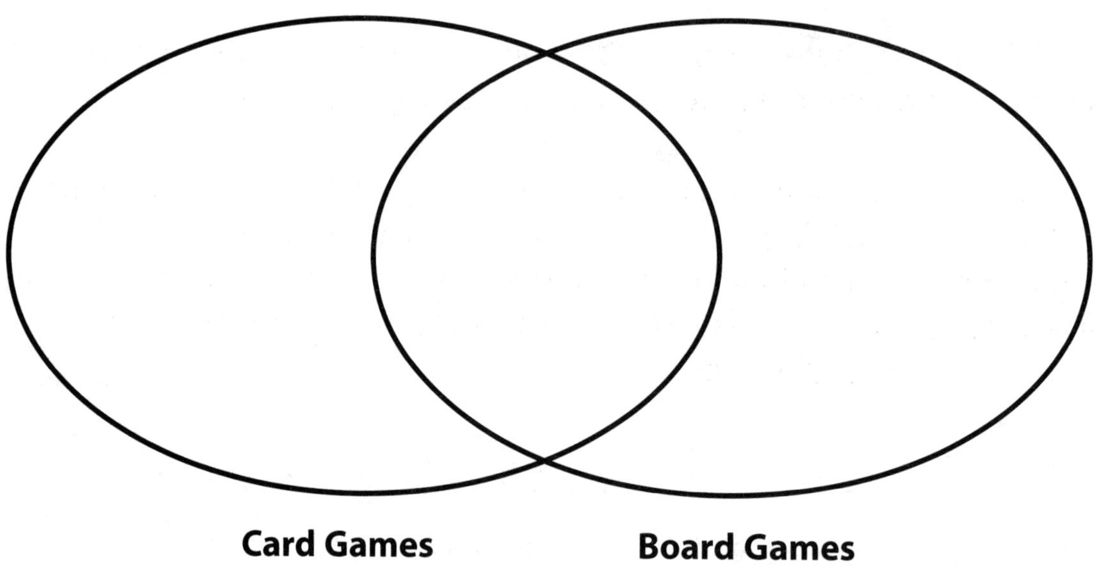

Name ______________________________ Date ______________

Vacations

SHOW ME THE WAY TO USE LOGIC SQUARES

Shiloh, Lorenzo, Ethan, and Bonnie are writing reports about their summer vacations. They either visited Hawaii, Disneyland, the Grand Canyon, or the Oregon coast. Use the clues to match each child to the correct vacation spot:

- Lorenzo went someplace with the vowels "a" and "i" in its name.
- Ethan went to a ranger program about rivers and erosion.
- Bonnie saw the ocean.
- Shiloh rode a roller coaster.

	Hawaii	Disneyland	Grand Canyon	Oregon coast
Shiloh				
Lorenzo				
Ethan				
Bonnie				

Strategic Steps

1. The first clue narrows Lorenzo's vacation spot to Hawaii or Disneyland. Mark "X" under Grand Canyon and Oregon coast on Lorenzo's row to eliminate those choices.

2. Ethan would have gone to a ranger program about rivers and erosion at the Grand Canyon. In Ethan's box under Grand Canyon, mark "Y" for "yes." In the other children's rows, mark "X" in the box for Grand Canyon. Mark "X" in the box for each of the other vacation spots for Ethan.

3. The third clue eliminates Disneyland as Bonnie's vacation spot. Mark "X" in the box for Disneyland on Bonnie's row to eliminate this spot for Bonnie.

4. Shiloh would have riden a roller coaster at Disneyland. Mark "Y" under Disneyland and "X" for the remaining vacation spots in Shiloh's row. Then mark "X" in the other children's boxes to eliminate Disneyland as their vacation spot. The only vacation spot left for Lorenzo is Hawaii. That means that Bonnie must have visited Oregon. Which vacation spot did each child visit?

Shiloh ______________________ Lorenzo ______________________

Ethan ______________________ Bonnie ______________________

Name ______________________________ Date ______________

Logic Squares

A Wendy, Anika, Masa, and Tom have all lived in other countries. These countries are Scotland, Italy, Australia, and France.

Here are some clues to help match each child with the country in which they lived:

- Tom lived in a country with a "c" in the name.
- The countries that Masa and Anika lived in both have three syllables.
- Anika lived on a different continent than Tom and Wendy.
- Wendy lived in a country that is on an island.

In which country did each child live?

Wendy ______________

Anika ______________

Masa ______________

Tom ______________

	Scotland	**Italy**	**Australia**	**France**
Wendy				
Anika				
Masa				
Tom				

B Anton loves hats. He has a baseball cap, visor, cowboy hat, and sombrero. The colors of his hats are purple, red, black, and white.

Here are some clues to help find out the color of each hat:

- The baseball cap was his favorite color, one of the colors on the United States flag.
- Anton has a hard time finding the cowboy hat in the darkness of his closet.
- The visor was blue until he dropped it into a bucket of red dye, now it is a new color.
- The sombrero gets dirty easier than the baseball cap because it's a lighter color.

Which color was each hat?

baseball cap ______________

visor ______________

cowboy hat ______________

sombrero ______________

	purple	**red**	**black**	**white**
baseball cap				
visor				
cowboy hat				
sombrero				

Name ______________________________ Date ______________

Summertime

A This summer, Decker, Brittany, Jenna, and Hannah all plan on leaving their homes in Kansas and visiting a national park for vacation. The parks are Yellowstone, Grand Canyon, Colonial, and Redwood national parks.

Here are some clues to help match each child to the park he or she will visit:

- Jenna is the only one traveling east.
- Brittany and Hannah are both visiting national parks whose names are compound words.
- The national parks that Brittany and Jenna are visiting are in states that touch an ocean.

Which park is each child going to visit?

Decker ______________________

Brittany ______________________

Jenna ______________________

Hannah ______________________

	Yellowstone	Grand Canyon	Colonial	Redwood
Decker				
Brittany				
Jenna				
Hannah				

B Alfonso, Christina, Mabel, and Becky were working together to plant a vegetable garden. Each child planted one vegetable—either carrots, beets, eggplant, or corn.

Here are some clues to help determine each child's vegetable:

- Christina's vegetable started with the same letter as her name.
- Becky's vegetable grows underground.
- Alfonso loved the red color of his vegetable.

Which vegetable did each child plant?

Alfonso ______________________

Christina ______________________

Mabel ______________________

Becky ______________________

	carrots	beets	eggplant	corn
Alfonso				
Christina				
Mabel				
Becky				

Answer Key

Homework Time (page 4)

4:05 p.m.

Backwards Time (page 5)

A. 9:40 a.m.

B. 7:15 a.m.

C. 5:55 p.m.

Hobbies (page 6)

A. 17 cookies

B. 4 miles

C. 4 books

Air Hockey Tournament (page 7)

6 games

Geography Bee (page 8)

15 questions

Game Time (page 9)

28 games

A Penny Saved (page 10)

A. $96.00

B. $24.00

The Pet Store (page 11)

A. 6 tetras

B. 21 cans

C. $18.00

Insects (page 12)

no; 17 more

Field Trip (page 13)

43 students

Rock Quarry (page 14)

A. 26 pounds

B. 13 students

The Fishing Trip (page 15)

A. 314 miles

B. 18 trout

Cardinals' Season (page 16)

A. 55 hits

B. 15 hits

Reading Contest (page 17)

A. 2,958 pages

B. 22 pages

C. 642 pages

Pet Parade (page 18)

A. Mr. Sutherland's

B. 1 more cat

C. 30 dogs

D. 7/30

Movie Treats (page 19)

soda and a candy bar

Make Your Best Guess (page 20)

A. Shannon — 12 pencils
Kate — 8 pencils

B. Tom — 43 cards
Gordon — 31 cards

C. $4.50 and $5.50

D. Becca — 20 minutes
Miranda — 15 minutes

School Olympics (page 21)

A. Roger — 4 feet
Xavier — 5 feet

B. Brenna — 8 seconds
Neha — 11 seconds

C. Edward — 10 feet
Tyler — 15 feet

D. Annabelle — 26 jumps
Mike — 17 jumps

Number Changing Box (page 22)

add 2

What Happened? (page 23)

A. subtract 16

B. multiply by 3

C. answers will vary

My Number Is Different! (page 24)

A. multiply by 2, add 1

B. multiply by 3, add 2

C. answers will vary

Flower Garden (page 25)

A. 48 flowers

B. 93 flowers

Pennies (page 26)

Total: $10.23

Day 1	1
Day 2	2
Day 3	4
Day 4	8
Day 5	16
Day 6	32
Day 7	64
Day 8	128
Day 9	256
Day 10	512

Pebbles (page 27)

Total: 1,275 pebbles

Day 1	5
Day 2	10
Day 3	20
Day 4	40
Day 5	80
Day 6	160
Day 7	320
Day 8	640

Lunch Line (page 28)

Ethan

Emma

Ben

Grant

Lauren

Back to School (page 29)

Room 8: Mia

Room 9: Lance

Room 10: Tara

Room 11: Jordan

Room 12: Nora

Science Notebooks (page 30)

Rita

Juan

David

Miguel

Susan

Leah

The Bridge (page 31)

1 person and 2 dogs (given) or 3 people and 1 dog

One Potato, Two Potatoes (page 32)

The following combinations will work:

5-pound bags	10-pound bags
0	5
2	4
4	3
6	2
8	1
10	0

The Picnic (page 33)

The following combinations will work:

People	Legs	Ants	Legs
9	18	1	6
6	12	2	12
3	6	3	18

Close Enough (page 34)

A. Answers may vary. Possible answer includes: You need to start over and count because you need an exact number to set a new record.

B. about 90 points

Estimation (page 35)

A. about 60 miles

B. Answers may vary. Possible answer includes: You cannot round the numbers. One time is faster than the other, and for a record you must be exact.

C. about 20 keychains

Friends (page 36)

A. about 120 cards

B. about 10:30 a.m.

C. Answers may vary. Possible answer includes: No, because if you round each number to the nearest ten, you would get 90–40=50. If you add 50 to 36, the number of days he has so far, you would get 86, which would not beat the record of 93 days.

Going to School (page 37)

3 miles; 15 miles

Paper Drive (page 38)

A. 250 pounds

B. 800 pounds

C. 4,000 pounds

Horseback Riding (page 39)

A. 35 pounds

B. 62 hours

C. 420 miles

Venn Diagram (page 40)

6 children

Play Time (page 41)

A. 3 students

B. 10 students

Survey Says (page 42)

A. 25 students

B. 13 children

Vacations (page 43)

Shiloh — Disneyland

Lorenzo — Hawaii

Ethan — Grand Canyon

Bonnie — Oregon coast

Logic Squares (page 44)

A. Wendy — Scotland

Anika — Australia

Masa — Italy

Tom — France

B. baseball cap — red

visor — purple

cowboy hat — black

sombrero — white

Summertime (page 45)

A. Decker — Grand Canyon

Brittany — Redwood

Jenna — Colonial

Hannah — Yellowstone

B. Alfonso — beets

Christina — corn

Mabel — eggplant

Becky — carrots